YOUR KNOWLEDGE HAS VALUE

- We will publish your bachelor's and master's thesis, essays and papers

- Your own eBook and book -
  sold worldwide in all relevant shops

- Earn money with each sale

Upload your text at www.GRIN.com
and publish for free

Bibliographic information published by the German National Library:

The German National Library lists this publication in the National Bibliography; detailed bibliographic data are available on the Internet at http://dnb.dnb.de .

Imprint:

Copyright © 2016 GRIN Verlag, Open Publishing GmbH
Print and binding: Books on Demand GmbH, Norderstedt Germany
ISBN: 9783668393233

Alfhonce Michael

Influenza H1N1 Virus as a Global Issue

GRIN Publishing

INFLUENZA H1N1 VIRUS AS A GLOBAL ISSUE

Over the years there has been a huge outbreak of the influenza H1N1 virus. It has caused stigma and fear among majority of the people all over the world. Little had been known about the disease before the year 1918 when the world was swept with multiple cases of the disease and approximately 500 million people getting affected and over 40 million people succumbed to the illness. The modern world came face to face with a huge pandemic in form of the influenza A H1N1 flu that affected numerous countries all over the world.

The influenza H1N1 virus is a respiratory disease that is highly infection and is known to majorly affect people who have come into recent contact with pigs or surfaces that have been contaminated by the swine flu viruses. The disease is known to display symptoms that are quite similar to those exhibited by people suffering from seasonal influenza. The H and the N in the name of the virus is extracted from the proteins that are found on the surface which are the glycoproteins; neuraminidase and haemagglutinin. These are what help differentiate the virus subtype from other influenza a subtypes. The virus is popularly known around the world as swine flu. The name was adopted because the disease had genes that upon undergoing several laboratory tests were found in pigs. The H1N1 flu virus is known to compose of genes that have been found in flu from various species including birds (avian flu), pigs as well as those found in flu viruses among humans.

The influenza H1N1 virus has undergone numerous mutations over a period of time to make it more adaptable to the human body and thus most medication that had been developed prior was almost obsolete. The flu is known to have jumped between species to be able to cause a condition in humans. The rapid spread of the virus in the year 2009 is attributed to the increased contact between human and human as well as the ease of travel from one area to another within a short period of time because of availability of air travel.

The H1N1 virus is transmitted from one person to another in the same manner as other influenza flus. It is known to enter the human body through the mouthy, nose, and the eyes. Therefore, one of the most common methods of transmission is coughing and sneezing. The germs released into the air is then transferred to others through breathing or may settle on surfaces such as tables, and doorknobs and contact with these surfaces causes an infection.

This is what ensured the rapid spread of the virus in the year 2009 when the disease was declared a global pandemic.

There is a huge history surrounding the disease. The flu was first discovered in the year 1918. The influenza flu had become well known in humans but was never known to affect pigs. From September 30th to October 5th, the world was hit with a huge pandemic that caused the death of about 40 to 50 million people caused by the influenza A H1N1 virus. The same year farmers especially in Iowa were puzzled by a disease that affected their pigs. The clinical syndromes of the swine disease had numerous similarities to the human disease developing interest in discovery of the causative agent.

There were numerous tests done on the virus between 1918 and 1947 where it was discovered that the virus rapidly mutates. From 1957 to 1977, there were no reported cases of the virus and it was thought to disappear. There is no clear reason to date why the virus disappeared. The strain that was found in 1977 was basically in the Soviet Union, Hong Kong and North-Eastern China. This strain was however the one that had been found in the 1950s and was thought to have been released from a laboratory by mistake.

Before 2009, there were few cases of the Influenza A H1N1 virus that had been reported and the cases brought forward were as a result of contact with pigs. In June 2009, the swine flu was elevated from an alert level of a pandemic alert level to a phase 6 alert level by the World Health Organization. This is the highest level that means that the disease has been reported in most of the countries in the world. The major reason for the rapid spread of the disease was the frequency of air travel and the human to human transmission. The next cases of swine flu that were reported were in the year 2015 and were found in India where it claimed 1371 lives. This was easier to control using the guidelines provided by the government to the affected states.

The virus can be spread one day before symptoms begin to show. The infected people can be contagious for as long as 7 days. Children below the age of 10 are able to pass on the disease for up to 10 days after they are infected. The symptoms of swine flu are diverse but just like any other flu they have symptoms that are quite similar. These symptoms include: runny or stuffy nose, chills, sneezing, dizziness, abdominal pains, and shortness of breath, watery eyes, confusion, no tears when crying, rash, irritability, bluish skin color, low temperature, fever, cough, sore throat, body aches, headache, and lack of appetite, fatigue, diarrhea and vomiting. There are numerous other complications that have been associated with the swine flu. Some of the complications include: severe pneumonia, lung infection as well as other breathing problems that may lead to death. One may also be prone diabetes and asthma.

In case one identifies the above symptoms, they should visit a doctor to get checked out since these symptoms may also be found with numerous other kinds of flus. The diagnosis should be done by a licensed doctor or medical practitioner. The doctor goes ahead to perform several laboratory tests to confirm the presence of the virus in one's bloodstream. The doctor should also find out if you have travelled to a part of the world where there have been recent reported cases of the H1N1 virus. There are various groups of people who should be tested for the virus. These are people who are in hospital and those who have a higher risk of contracting life-threatening problems from swine flu.

The people who are most prone to swine flu include: children under 5 years old, children and teenagers under the age of 18 who are undertaking aspirin therapy which is long-term and may expose them to the risk of Reyes' syndrome after exposure to swine flu, adults and children who have metabolic, neuromuscular, heart, lung, blood, nervous systems, and chronic liver problems and suppressed immune systems, people older than 65 years, pregnant women and people in long term care facilities.

There are numerous myths and misconceptions that have surrounded the influenza H1N1 virus. One of the most common myths about the virus is that if one is healthy, they do not need to worry about the disease. Even though there have been cases of people with the H1N1 fully recovering without having undergone medical treatment. There is no indication, however that good health protec6ts them from the serious complications associated with the virus. Another common one is that one can contact the illness by coming into contact with pork products. It is highly unlikely for one to get infected if they consume well cooked pork products that have been cooked to a temperature above 71 degrees Celsius or 160 degrees Fahrenheit.There is also the misconception that since the virus is inevitable, they should get over it and thus deliberately get infected. They think that infecting themselves will give them a better chance of controlling it. This should be avoided because the virus has different effects on different people and due to its mutative nature it may become dangerous in the future. The first 24 hours after contracting the illness, one may not even know they have the disease. They should however know that they are still contagious and should practice good hygiene habits.

Even though the H1N1 vaccine is administered in a bid to reduce the risk of infection, it is not a guarantee that one will not be infected. The great news is that there will be a lower chance of getting severe complications. The hugest misconception is that the vaccine can give you the flu. This is because some of the side effects of getting the flu shot include a bad cold, a slight fever and achiness which are also known symptoms of the H1N1 virus. The H1N1 flu is caused by a virus. Therefore those who decide to take antibiotics should understand that those are meant for bacteria and thus are not of much help. The antibiotics will not help and may have adverse side effects and a misuse may cause resistance in the future.

The countries that have been affected by the H1N1 influenza over the years have been numerous. In the 2009 pandemic, Mexico was the first country to report cases of the disease. Within one month the virus had spread to 32 states in the country and before long, cases were reported in the neighboring southern California in the United States. By April 2009, the swine flu was declared a public health emergency in the United States. In June of the same year, the World Health Organization had become a worldwide pandemic having spread from North America to other countries like the United Kingdom, Argentina, Australia, Chile, Japan, India and Spain. One of the most recent instances of the disease was India in 2015 which caused the death of over 1,842 people, the Nepal outbreak in 2015 that caused the death of 25 people and the 2016 Pakistan outbreak where there were 7 reported cases.

The disease is treated using antiviral drugs majorly used to treat seasonal flu. Some of these drugs may include: neuraminidase inhibitors such as peramivir(Rapivab), zanamivir(Relenza) and oseltamivir(Tamiflu) and M2 inhibitors such as amantadine. These are known as antiviral medication. Pharmacists and doctors advise that one should take the drugs within the first 48 hours of the first symptoms of flu so they can work better. One is strongly advised to avoid the use of antibiotics since this is caused by a virus not by bacteria. There have been discoveries that medications that are purchased over the counter as pain remedies, for the cold and the flu help relieve some of the symptoms.

In order to manage the spread of the disease, those who are sick should follow the following guidelines;

- They should avoid public areas and are strongly advised to stay at home. They should not attend school or go to work.
- They should rest and drink plenty of fluids.
- Should be quarantined and ensure they are at least a meter away from other people.

- To avoid the further spread of the disease, cover the mouth with a tissue, hands or your sleeves while sneezing or coughing. The used tissue should be carefully disposed while one must wash their hands thoroughly.
- One must regularly wash their hands thoroughly using soap and water for at least 15 seconds.

There are several measures that one can take to ensure that you do not contract the H1N1 flu virus including:

- Ensure that you do not come in contact with infected persons or those displaying symptoms of the H1N1 flu.
- Thorough and regular washing of the hands with soap and clean water to ensure proper sanitation reduces the risk of exposure to the virus. Hands should be washed for a minimum of 15 seconds. In cases where water and soap is unavailable, one should use an alcohol-based hand sanitizer.
- Good health habits should be practiced including a good night sleep, a well-balanced diet and physically fit.
- Avoid touching your eyes, mouth or nose if suspicious that you have been exposed to a person with the flu.

A pandemic's degree is defined by the severity of the disease. It has not however been determined whether this is characterized by the number of illnesses or deaths caused. The influenza H1N1 was at its worst when it caused the deaths of approximately 50 million people between the year of 1918 and 1919. There have been cases of the virus that have sprung up in the past year sporadically among humans.

The most recent global pandemic was in 2009 which was caused by the influenza A (H1N1) virus. It was earlier detected in North America. The United States soon declared it as a national public health emergency. It was of huge international concern because there were numerous gaps in the health capacity both globally and locally concerning the virus, limited scientific knowledge about the virus, uncertainty in decision making, international cooperation was a complex issue and communication between the public, policy makers and the experts.

The emergence of the influenza H1N1 virus in the United States came as surprise to most medical experts since it was expected to emerge in Asia. It spread at an alarming rate to the rest of the world due to the efficiency of international travel and commerce. This pandemic brought forward the fact that there has been a rise in the cases of diseases that have occurred cross species. It showed that there has been a higher involvement of animals in the emergence of new diseases, their spread and their establishment.

There has been a vaccine that has been manufactured since then to help curb the effects and the spread of the influenza A(H1N1) virus. There are two types of the vaccine; the nasal spray flu vaccine or the flu shot which is administered via the arm through a needle. There was a huge wave of doubt that accompanied the announcement of the development of the H1N1 swine flu vaccine. This was mainly due to the effects of the 1976 swine flu vaccine that caused the development of the Guildin-Barre syndrome. It should be noted that no vaccine is 100% safe and therefore those with egg allergies are not advised to use it since they are used in its manufacture. There is also a chance that one in 1 million of people who use the vaccine are likely to develop the Guidin-Barre Syndrome (GBS).

In the past, there have been several outbreaks of the influenza pandemic within each decade. There is no way to predict the severity or the danger that these outbreaks may have on human health. There is a necessity therefore for the development of better epidemiological understanding as well as deeper biological research on the virus to help reduce this uncertainty. There has been an improvement in the disease detection technique now used in most countries, the laboratory capacity has been considerably expanded and there is availability of superior surveillance to catch emerging issues surrounding the virus. Most countries have agreed to vaccine sharing to ensure the spread of the disease is curbed at its early stages. There are new technologies that have been developed to come up with better antiviral agents, a production capacity that is greater, faster throughput and influenza vaccines that are more effective. The methods used in the past to produce influenza vaccines were based on egg cultures and have been proven to work too slowly against the rapidly mutating influenza virus.

The virus came upon the world by surprise, there have been several conferences held by the World Health organization to ensure that in the future the pandemic does not catch the world by surprise again and that it is controlled as soon as possible. Most medical experts have expanded and deepened the research surrounding the virus. There have been projections and speculations on the mutations that the influenza H1N1 virus may undergo in the future.

The governments and other global health institutions have invested heavily into the business of ensuring that the virus has been controlled and vaccines have been developed to immunize people.

In the future, there seems to be a higher chance for such a pandemic to be caught before the effects is spread far and wide to prevent the adverse effects it may have on families, the community, countries and the world at large. Everyone is advised to always be aware of their surrounding and ensure that they maintain good health practices to keep at bay all infections. The governments of all nations are also advised to ensure that their health facilities are up to standards that can control the spread and reinfection of the infected.

References

Bartolotti, C. (2010). The H1N1 influenza pandemic of 2009 (1st ed.). New York: Nova Science Publishers.

Canada's response to the 2009 H1N1 influenza pandemic. (2010) (1st ed.). Ottawa, Ont.

Obama, B. (2009). Declaration of a national emergency with respect to the 2009 H1N1 influenza pandemic in the United States (1st ed.). Washington: U.S. G.P.O.

Stroud, C., Nadig, L., & Altevogt, B. (2010). The 2009 H1N1 influenza vaccination campaign (1st ed.). Washington, D.C.: National Academies Press.

YOUR KNOWLEDGE HAS VALUE

- We will publish your bachelor's and
  master's thesis, essays and papers

- Your own eBook and book -
  sold worldwide in all relevant shops

- Earn money with each sale

Upload your text at www.GRIN.com
and publish for free